In the
Kitchen

by Dorothy Heil

Contents

3

solid

A **solid** has its own shape.

This pot is a **solid.**

liquid

A **liquid** takes the shape of its container.

This sauce is a **liquid.**

weight

Weight is how heavy or light something is.

The napkins have a lighter **weight** than the plates.

temperature

Temperature is how hot or cold something is.

This pasta has a hot **temperature.**

float

When something **floats,** it stays on or near the top of a liquid.

This spoon **floats** in the water.

sink

When something **sinks,** it falls to the bottom of a liquid.

float

liquid

property

sink

solid

temperature

texture

weight

This plate **sinks** to the bottom of the water.

Properties in the Kitchen

It's time to eat! The people in this family use their senses when they make pasta and salad for dinner.

Their senses tell them about **properties.** How salad and pasta taste, look, feel, and smell are all properties.

property

A **property** is something about an object that you can observe with your senses.

The herbs, spices, and sour cream in this kitchen have different colors. Color is a property.

These ingredients are red, green, and white.

The boy mixes spices into the salad dressing. A property of the dressing changes. Now it has a new color.

These salad vegetables are different sizes.

The lettuce is bigger than the peppers.

Size is a property.

The man cuts a tomato for the salad.
Cutting changes a property of the tomato.
Now he has smaller pieces.

The man covers the salad with foil. He bends the foil over the bowl. Now the foil has a new shape.

Next, he chooses some salad plates. He can choose round or square plates. Shape is also a property.

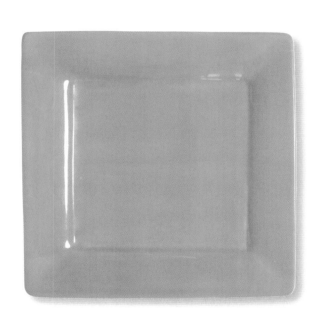

The properties of pasta can be changed in different ways. The woman breaks the hard pasta. Breaking the pasta changes its size.

Breaking pasta is a small change.

Pasta is hard before you cook it. It is soft after you cook it in hot water.

Cooking pasta in water is a big change.

The boy wears an apron while he helps with dinner. The apron has a soft **texture.**

texture

Texture is the way an object feels.

He grates cheese for the pasta. This cheese grater has a rough texture. Texture is how something feels.

Texture is a property.

Solids and Liquids

The woman stirs sauce in a pot. The pot is a **solid.** It has its own shape.

solid

A **solid** has its own shape.

The sauce is a **liquid.** It takes the shape of the pot and the ladle.

liquid

A **liquid** takes the shape of its container.

Comparing Properties

The boy gets plates and napkins from
the cabinet.

The plates have a heavier **weight** than the napkins. Weight is a property of the plates and the napkins.

weight

Weight is how heavy or light something is.

The foods at the dinner table have different **temperatures.** The pasta is hot. The salad is cold.

temperature

Temperature is how hot or cold something is.

The ice water is also cold. Temperature is a property.

Dinner is over. It is time to wash the dishes. The wooden spoon **floats** in the water.

What other objects float or sink in the water?

float

When something **floats,** it stays on or near the top of a liquid.

The forks, knives, and cheese grater **sink.**
Whether something floats or sinks is a property.

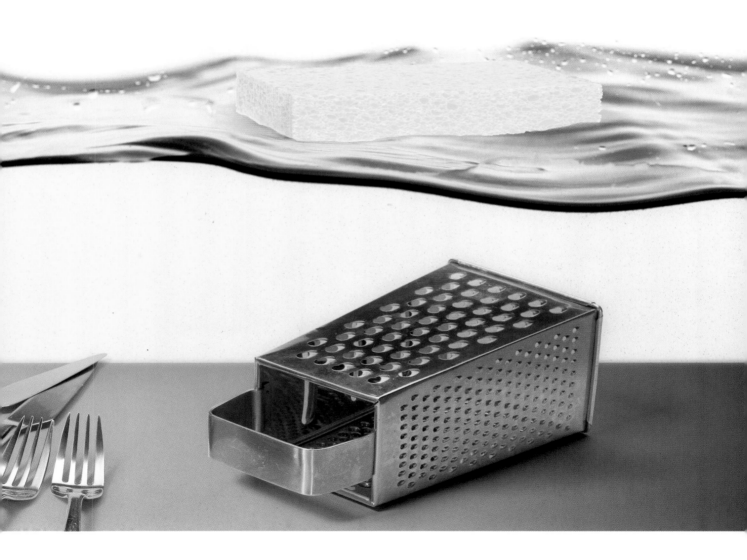

sink

When something **sinks,** it falls to
the bottom of a liquid.

Conclusion

The family chose foods to make based on their properties. They changed the foods by breaking, cutting, and cooking. The family used solids and liquids to make the salad and pasta.

Think About the Big Ideas

1. How can you describe the properties of a tomato?
2. How can you compare the properties of a napkin and a plate?
3. What are some solids and liquids you use when you are cooking?

Share and Compare

Turn and Talk

Compare the properties of the objects shown in your books. How are they alike or different?

Read

Find a photo with a caption. Read the caption to a classmate.

Write

Describe how you can change properties of an object. Share what you wrote with a classmate.

Draw

Show objects with different properties in your classroom. Share your drawing with a classmate.

Meet Stephon Alexander

Scientists ask questions and share their ideas. Stephon Alexander is a scientist who asks questions about the properties of space.

His students also ask him questions that give him new ideas. Together, they study the properties of space.

Index

Acknowledgments
Grateful acknowledgment is given to the authors, artists, photographers, museums, publishers, and agents for permission to reprint copyrighted material. Every effort has been made to secure the appropriate permission. If any omissions have been made or if corrections are required, please contact the Publisher.

Photographic Credits:
Cover (bg) Piotr Rzeszutek/Shutterstock; Cvr Flap (t, c, b), 2-28 Mark Thiessen and Becky Hale, National Geographic Photographers; Title (bg) Anthony Blake/Fresh Food Images/Photolibrary; 26-27 (bg) Brian Hagiwara/Brand X Pictures/Jupiterimages; 31 Mark Thiessen/National Geographic Image Collection; Inside Back Cover (bg) Tomo Jesenicnik/ Shutterstock.

Neither the Publisher nor the authors shall be liable for any damage that may be caused or sustained or result from conducting any of the activities in this publication without specifically following instructions, undertaking the activities without proper supervision, or failing to comply with the cautions contained herein.

Published by National Geographic School Publishing & Hampton-Brown
Sheron Long, Chairman
Samuel Gesumaria, Vice-Chairman
Alison Wagner, President and CEO
Susan Schaffrath, Executive Vice President, Product Development

Editorial: Fawn Bailey, Joseph Baron, Carl Benoit, Francis Downey, Richard Easby, Mary Clare Goller, Chris Jaeggi, Carol Kotlarczyk, Kathleen Lally, Henry Layne, Allison Lim, Taunya Nesin, Paul Osborn, Chris Siegel, Sara Turner, Lara Winegar, Barbara Wood

Art, Design, and Production: Andrea Cockrum, Kim Cockrum, Adriana Cordero, Darius Detwiler, Alicia DiPiero, David Dumo, Jean Elam, Jeri Gibson, Shanin Glenn, Raymond Godfrey, Raymond Hoffmeyer, Rick Holcomb, Cynthia Lee, Anna Matras, Gordon McAlpin, Melina Meltzer, Rick Morrison, Cindy Olson, Christiana Overman, Andrea Pastrano-Tamez, Sean Philpotts, Leonard Pierce, Cathy Revers, Stephanie Rice, Christopher Roy, Janet Sandbach, Susan Scheuer, Margaret Sidlosky, Jonni Stains, Shane Tackett, Andrea Thompson, Andrea Troxel, Ana Vela, Teri Wilson, Brown Publishing Network, Chaos Factory, Inc., Feldman and Associates, Inc.

The National Geographic Society
John M. Fahey, Jr., President & Chief Executive Officer
Gilbert M. Grosvenor, Chairman of the Board

Manufacturing and Quality Management,
The National Geographic Society
Christoper A. Liedel, Chief Financial Officer
George Bounelis, Vice President

Copyright © 2010 The Hampton-Brown Company, Inc., a wholly owned subsidiary of the National Geographic Society, publishing under the imprints National Geographic School Publishing and Hampton-Brown.

All rights reserved. No part of this book may be reproduced or transmitted in any form or by any means, electronic or mechanical, including photocopying, recording, or by an information storage and retrieval system, without permission in writing from the Publisher.

National Geographic and the Yellow Border are registered trademarks of the National Geographic Society.

National Geographic School Publishing
Hampton-Brown
P.O. Box 223220
Carmel, California 93922
www.NGSP.com

Printed in the USA.

ISBN: 978-0-7362-5539-4

10 11 12 13 14 15 16 17

10 9 8 7 6 5 4 3 2 1